THE POETRY OF CESIUM

The Poetry of Cesium

Walter the Educator

SKB

Silent King Books a WhichHead Imprint

Copyright © 2023 by Walter the Educator

All rights reserved. No part of this book may be reproduced in any manner whatsoever without written permission except in the case of brief quotations embodied in critical articles and reviews.

First Printing, 2023

Disclaimer
This book is a literary work; poems are not about specific persons, locations, situations, and/or circumstances unless mentioned in a historical context. This book is for entertainment and informational purposes only. The author and publisher offer this information without warranties expressed or implied. No matter the grounds, neither the author nor the publisher will be accountable for any losses, injuries, or other damages caused by the reader's use of this book. The use of this book acknowledges an understanding and acceptance of this disclaimer.

"Earning a degree in chemistry changed my life!"
– Walter the Educator

dedicated to all the chemistry lovers, like myself, across the world

CONTENTS

Dedication v

Why I Created This Book? 1

One - Beauty Of Science 2

Two - Rare Element 4

Three - Cesium, We Salute You 6

Four - Captivating Us All 8

Five - World Of Science 10

Six - Wonders Start 12

Seven - Testament 14

Eight - Science's Embrace 16

Nine - Own Shine 18

Ten - Shining Star 20

Eleven - Unfolds 22

Twelve - Cesium, The Element 24

Thirteen - Cesium, The Light 26

Fourteen - Master Of Time 28

Fifteen - Shines So Bright 30

Sixteen - Symbol Of Progress 32

Seventeen - Cesium's Beauty 34

Eighteen - Vivid Dream 36

Nineteen - Forever Sublime 38

Twenty - Inspiring 40

Twenty-One - Knowledge 42

Twenty-Two - Scientific Love 44

Twenty-Three - Flawless Law 46

Twenty-Four - Expanding Horizons 48

Twenty-Five - Greatest Treasure 50

Twenty-Six - Day By Day 52

Twenty-Seven - Curiosity Resounds 54

Twenty-Eight - Minds Have Grown 56

Twenty-Nine - Unraveling Mysteries 58

Thirty - Forever In Our Hearts 60

Thirty-One - Innovation's Partner 62

Thirty-Two - Darkest Night 64

Thirty-Three - Cesium, The Element 66

Thirty-Four - Scientific Flight 68

Thirty-Five - Poetic Delight 70

About The Author 72

WHY I CREATED THIS BOOK?

Creating a poetry book about the chemical element of Cesium was a unique and intriguing project. It allows for the exploration of the element's properties and characteristics in a creative and artistic way. Poetry has the power to evoke emotions and connect with readers on a deeper level. By incorporating Cesium into poetry, I can create a metaphorical language that captures the essence of the element and its significance. This book can serve as a bridge between science and art, engaging readers in a thought-provoking and imaginative journey through the world of Cesium.

ONE

BEAUTY OF SCIENCE

In the realm of elements, a jewel gleams,
A metal so rare, Cesium it seems.
With atomic number 55, it stands tall,
A marvel of nature, fascinating all.

    A silvery metal, soft to the touch,
Cesium, the element that can do so much.
Its beauty belies a fiery soul,
A dance of electrons, a story yet untold.

    In the depths of space, where stars collide,
Cesium's secrets, they cannot hide.
It glows with a hue, a vibrant blue,
Unveiling its power, to me and to you.

    With a single electron, so easily lost,
Cesium's reactivity, a tempest tossed.

It ignites with a spark, a flicker of flame,
A chemistry enchantment, a dangerous game.

From atomic clocks to photoelectric cells,
Cesium's properties, a science that compels.
It measures time with precision untold,
A testament to its worth, a story to be told.

Oh, Cesium, the element of grace,
Your mysteries we cannot erase.
In laboratories and distant skies,
Your brilliance continues to mesmerize.

So, let us marvel at Cesium's might,
A radiant element, shining so bright.
For in its essence, we find delight,
In the beauty of science, day and night.

TWO

RARE ELEMENT

In realms of science, a wondrous find,
A metal rare, just one of its kind,
Cesium, the element, atomic fifty-five,
With silvery sheen, it comes alive.
    A dance of electrons, so swift and bright,
Reactivity, its claim to delight,
In nature's realm, it seeks its place,
A metal that leaves no trace.
    Atomic clocks, they owe their might,
To Cesium's oscillations, pure and right,
With precision, it ticks away,
Marking time, night and day.
    Photoelectric cells, they harness its power,
Transforming light, hour by hour,

Cesium's electrons, they leap and play,
Captivating observers, who gaze away.

    Oh, Cesium, you beauty rare,
In labs, scientists, they truly care,
For your grace, your elegance, so fine,
You make their hearts and minds entwine.

    From the depths of Earth, you rise above,
A metal that fills our hearts with love,
Cesium, dear friend, we sing your praise,
For the wonders you bring, in countless ways.

    So, let us celebrate this metal divine,
In science's realm, a treasure we find,
Cesium, you captivate, you inspire,
A rare element, that sets our hearts on fire.

THREE

CESIUM, WE SALUTE YOU

In the realm of atoms, a jewel does reside,
A shimmering element, with beauty inside.
Cesium, the name that graces its form,
A dancer of electrons, in the atomic storm.

With a silvery sheen, it shines so bright,
A metal soft and malleable, a captivating sight.
Its brilliance does beckon, a mesmerizing hue,
Cesium, oh Cesium, how we adore you.

In atomic clocks, you find your place,
With precision and grace, you set the pace.
Ticking away, counting the seconds in time,
Cesium, the master, in this cosmic rhyme.

Photoelectric cells, your power revealed,
Harnessing the light, as energy is sealed.

From sunlight to electricity, a magical connection,
Cesium, the conductor, of this energy's perfection.
   But beware, dear Cesium, for you hold within,
A reactivity that can make heads spin.
With water, a dance of fire you perform,
A dazzling display, a true chemical storm.
   Oh Cesium, you captivate, you enthrall,
A beauty in science, you stand tall.
From your atomic wonders to your fiery might,
Cesium, we salute you, shining so bright.

FOUR

CAPTIVATING US ALL

In the realm of elements, a jewel does reside,
A radiant metal, with its glow, it does confide,
Cesium, the name that echoes through the land,
A rare beauty, held in nature's loving hand.

Oh, Cesium, so reactive, so quick to ignite,
In water's embrace, you dance with pure delight,
Bursting forth with energy, a fiery display,
A symphony of colors, as you go on your way.

Atomic clocks, they owe their precision to you,
Ticking away the moments, accurate and true,
With your synchronized heartbeat, time is defined,
In perfect harmony, the world is aligned.

Photoelectric cells, where your magic is revealed,
Capturing the essence of light, your power unconcealed,

Transforming photons into energy, you shine,
A catalyst of progress, a marvel so divine.

Cesium, you are a muse, inspiring in every way,
Scientists and dreamers, in awe of you they sway,
Your allure is captivating, a magnet to the soul,
Filling hearts with love, making them whole.

So here's to you, Cesium, with your brilliance untold,
A beacon in the darkness, a treasure to behold,
In the realm of elements, you stand so tall,
Enthralling minds and hearts, captivating us all.

FIVE

WORLD OF SCIENCE

In the realm of elements, a beauty divine,
Cesium, a treasure, oh how it does shine.
Softly it glimmers, a metallic embrace,
With a golden hue, it lights up the space.

Reactive it is, with a fiery zeal,
Eager to bond, its desires reveal.
With water it dances, a sizzling show,
Igniting the air, a spectacular glow.

In atomic clocks, it finds its domain,
Ticking away, in precision it reigns.
A measure of time, a constant heartbeat,
Cesium's rhythm, so flawless and neat.

Photoelectric cells, they harness its might,
Capturing photons, converting their light.

Efficient and powerful, it leads the way,
Illuminating the world, night and day.

Oh Cesium, enchanting, a scientific muse,
Captivating minds, with grace it infuses.
A symbol of elegance, a scientific gem,
In laboratories, it shines like a diadem.

From reactivity to precision sublime,
Cesium, you've etched your mark, for all time.
A muse to inspire, a wonder to behold,
In the world of science, your story is told.

SIX

WONDERS START

In the realm of elements, a jewel does reside,
A shimmering beauty, with secrets to confide.
Cesium, the name that captures the eye,
Reactive and rare, against the cerulean sky.

With fiery passion, it dances in the air,
Igniting with water, a spectacle so rare.
Its atoms aglow, like a celestial show,
Cesium's radiance, a sight to behold.

In the heart of atomic clocks, it does reside,
Ticking with precision, where time does confide.
Its electrons pulsate, in perfect harmony,
Counting the seconds, with unwavering accuracy.

In photoelectric cells, it finds its domain,
Harnessing light, like a celestial crane.

Absorbing the photons, it transforms their might,
Unleashing energy, in a symphony of light.

Oh Cesium, a treasure, both rare and refined,
In science's embrace, your wonders we find.
A dancer of elements, with beauty untold,
Cesium, your story forever unfolds.

So let us revel, in your captivating glow,
Illuminate our minds, like the stars that bestow.
Cesium, the element, that enchants the heart,
A symphony of science, where wonders start.

SEVEN

TESTAMENT

In the realm of atoms, a jewel holds its sway,
A metal so rare, that dances with the day.
Cesium, the element, a radiant sight,
Unveils its secrets, with ethereal might.

    In photoelectric cells, where dreams take flight,
Cesium's allure, a mesmerizing light.
For when the photons, with grace, cascade,
Cesium's electrons, they gleefully invade.

    A dance of energy, a symphony of rays,
Cesium, the conductor, orchestrating the plays.
From silver to gold, it effortlessly transforms,
In the presence of light, its beauty reforms.

    But lo, another tale, of precision and time,
Cesium, the maestro, with a rhythm so sublime.

In atomic clocks, where seconds are divine,
Cesium's heartbeat, a metronome so fine.
   With pulsating frequency, it measures the tick,
Cesium's vibrations, never missing a click.
A beacon of accuracy, a celestial guide,
Cesium, the timekeeper, in the universe wide.
   Oh, Cesium, element of allure and grace,
Your reactivity, a charm we can't erase.
From laboratories to stars, you leave your mark,
A catalyst for knowledge, forever in the dark.
   So let us celebrate this element rare,
Cesium, the enchantress, beyond compare.
In science's embrace, you forever reside,
A testament to the wonders that nature provides.

EIGHT

SCIENCE'S EMBRACE

In the realm of atoms, a marvel resides,
A metal so rare, Cesium it's called,
With electrons dancing in valence's stride,
Its beauty and brilliance, a sight to behold.

In atomic clocks, its purpose is grand,
Ticking away with precision untold,
Measuring time with a delicate hand,
As seconds unfold, its secrets unfold.

With every tick and every tock,
Cesium marks the passing of time,
A constant companion on life's winding clock,
An ally in rhythm, sublime and prime.

And yet, Cesium's wonders don't cease,
For in photoelectric cells it does shine,

Harnessing light and bestowing release,
Its power, a gift, so divine.
    Its reactivity, a scientific treasure,
Unveiling secrets, unlocking the unknown,
In laboratories, it's a priceless measure,
A catalyst for knowledge, forever sown.
    Oh, Cesium, element of awe and might,
Your properties, a symphony of grace,
Through time and light, you guide our sight,
In science's embrace, you claim your place.

NINE

OWN SHINE

In the heart of atomic clocks, you reside,
Cesium, the element that marks the stride,
With gentle oscillations, precise and true,
You measure time, unveiling moments anew.
    In photoelectric cells, your luminescence gleams,
Harvesting light, fulfilling dreams,
Your electrons dance, a radiant display,
Harnessing energy, guiding the way.
    Oh, Cesium, you're a beacon of time,
A master of rhythm, a metronome sublime,
With atomic precision, you never stray,
Ticking away the seconds, night and day.
    Your beauty lies in your reactivity,
A shimmering metal, full of activity,

Igniting with a flame, a vibrant hue,
Unleashing your secrets, revealing what's true.
    Cesium, you enchant with your vibrant glow,
A catalyst for progress, a curiosity to know,
From atomic clocks to scientific quests,
You mark the passage of time, unlocking life's tests.
    Oh, Cesium, element of allure,
With every tick, you endure,
A testament to nature's grand design,
In your atomic embrace, we find our own shine.

TEN

SHINING STAR

In the realm of atoms, a shining star,
Cesium, the element that travels far.
With electrons dancing in their orbits wide,
A tale of precision it does confide.

Atomic clocks, where time takes its cue,
Cesium, the heartbeat, so steady and true.
It oscillates, resonates, a rhythmic dance,
Measuring moments with utmost advance.

In photoelectric cells, it finds its place,
Absorbing light with its radiant grace.
Electrons leap, energized by its touch,
Harnessing photons, oh, it means so much.

A unique poem, you ask, I shall create,
Another facet of Cesium we shall celebrate.

A timekeeper it is, with unrivaled might,
Counting the seconds, from day to night.

Its reactivity, a quality profound,
In labs, experiments, it can be found.
With a sizzle and a spark, it ignites the air,
Unleashing its power, a chemist's affair.

And now, one more poem, I shall devise,
To marvel at Cesium, as it mesmerizes.
Guiding energy, with a shimmering gleam,
A metal so rare, like a dream within a dream.

Cesium, a marvel, enchanting to behold,
A symphony of elements, a story yet untold.
In its depths, mysteries and wonders lie,
A shining star, lighting up the sky.

ELEVEN

UNFOLDS

In the realm of atoms, where time takes its toll,
There lies a secret, a tale yet untold.
A metal so rare, a treasure untamed,
Cesium, the element, with glory untamed.

In atomic clocks, it holds the key,
Precision and rhythm, for all to see.
Its electrons dance, in perfect embrace,
Keeping time's heartbeat, with elegant grace.

A beacon of light, it harnesses with might,
Shining through darkness, illuminating the night.
Its glow, a reminder, of its power untamed,
Cesium, the element, forever acclaimed.

But beyond the clocks, a secret it hides,
A reactivity, that science abides.

With a touch so gentle, it ignites with a spark,
A dance of fire, in the deep and the dark.
  In labs it's studied, in experiments grand,
Unleashing its energy, with a wave of its hand.
A catalyst rare, in reactions unseen,
Cesium, the element, a scientific dream.
  So let us marvel, at this element divine,
With its timekeeping prowess, and reactivity so fine.
Cesium, the element, a wonder to behold,
In the world of atoms, its story unfolds.

TWELVE

CESIUM, THE ELEMENT

In the realm of atoms, where secrets unfold,
There lies a treasure, a story yet untold.
Cesium, the timekeeper, ruler of the clock,
Unleashing energy, like a celestial rock.

With electrons dancing, in orbits they trace,
Cesium's allure, none can efface.
A metal so gentle, yet fiercely reactive,
Its mysteries and wonders, truly captivate.

In atomic clocks, its purpose takes flight,
Ticking away, with precision so tight.
Counting the seconds, the minutes, the hours,
Cesium's guidance, like magical powers.

In laboratories, its knowledge unearths,
Catalyst for breakthroughs, revealing its worth.

Unlocking the secrets, in test tubes it gleams,
Cesium, the element, of scientific dreams.

    Its vibrant blue flame, a mesmerizing sight,
Illuminating the darkness, with purest light.
A beacon of discovery, shining so bright,
Cesium, the element, guiding our sight.

    Oh, Cesium, element of grace,
With your atomic weight, a cosmic embrace.
From the depths of the Earth, to the heavens above,
You remind us of wonders, and the power of love.

    So let us celebrate, this element divine,
For Cesium's beauty, forever shall shine.
A testament to science, and human endeavor,
Cesium, the element, we shall cherish forever.

THIRTEEN

CESIUM, THE LIGHT

In the realm of atoms, a shimmering glow,
A secret lies, that few truly know.
Cesium, the element, with power untold,
A catalyst for progress, a story to unfold.

With ticking precision, it keeps time's beat,
A timekeeper supreme, never to retreat.
Its electrons dance, in orbits so grand,
Revealing the secrets of our vast land.

Reactive and bold, in the lab it's adored,
Unleashing reactions, like never before.
Igniting the flames of scientific might,
Cesium, the spark, that illuminates the night.

In the hands of the curious, it reveals,
The mysteries of nature, its hidden appeals.

Its atomic structure, a marvel to see,
Cesium, the beauty, that sets atoms free.

Oh, Cesium, precious element divine,
With allure and grace, you forever shine.
Unlocking the secrets, unraveling the unknown,
In your atomic dance, the universe is shown.

A gem of the periodic table, so rare,
With worth beyond measure, beyond compare.
Guiding our journey, through darkness it gleams,
Cesium, the light, in the realm of dreams.

FOURTEEN

MASTER OF TIME

In the realm of atoms, a celestial dancer,
Cesium takes its place, a captivating enhancer.
A timekeeper, it counts the seconds with grace,
Unveiling mysteries, through its atomic embrace.

A metal so rare, with a golden hue,
Cesium shines bright, in shades of cobalt blue.
Its electrons dance, in quantum delight,
Unveiling the secrets, of the cosmic night.

Reactive and fierce, it ignites with a flame,
Caressing the darkness, with a radiant name.
From the depths of the Earth, it emerges with might,
A catalyst for science, a beacon of light.

In laboratories, it reveals the unknown,
Unraveling the secrets, that nature has sown.
From atomic clocks, to lasers so grand,
Cesium's power, we can barely understand.

Oh, Cesium, you hold the universe's key,
Unleashing the wonders, for all eyes to see.
A symbol of beauty, a spark of pure fire,
You inspire the dreamers, fueling their desire.

So let us raise a toast, to this elemental star,
Cesium, a treasure, no matter how far.
In its atomic realm, it dances and gleams,
A master of time, unlocking our dreams.

FIFTEEN

SHINES SO BRIGHT

In the realm of time, where moments unfold,
There lies a secret, untold, untold.
Cesium, the keeper of the ticking clock,
Its atomic rhythm, a cosmic lock.

Reactive and wild, it dances in air,
A fiery passion, beyond compare.
With flames of violet, it blazes bright,
Igniting the darkness, a radiant light.

In the lab of science, it takes its stage,
Unleashing its power, with boundless rage.
From atomic clocks, precise and true,
To tests of reactivity, it will pursue.

Oh, Cesium, you hold the key,
To unlocking secrets, for all to see.

Through your properties, we understand,
The wonders of chemistry, hand in hand.

A shining star, in a world so vast,
Your beauty, like a spell, forever cast.
Guiding our progress, igniting our dreams,
Cesium, the element that gleams.

So let us marvel at your brilliant hue,
And cherish the lessons you help us accrue.
For in your presence, we find delight,
Cesium, the element that shines so bright.

SIXTEEN

SYMBOL OF PROGRESS

In the realm of atoms, a marvel we find,
A radiant element, one of its kind.
Cesium, the timekeeper, ticking away,
Unveiling secrets, day after day.

A catalyst it is, a spark in the dark,
Igniting reactions, leaving its mark.
With reactivity fierce, it dances with ease,
Unleashing its power, a force to appease.

A source of energy, vibrant and bright,
Its flame burns blue, illuminating the night.
A beacon of light, in darkness it gleams,
Guiding the way through scientific dreams.

In labs and experiments, it takes the stage,
Unraveling mysteries, turning the page.

A symbol of discovery, it stands tall,
Unveiling the wonders, revealing them all.

    Its beauty unmatched, a sight to behold,
With shimmering hues, like liquid gold.
Cesium, oh Cesium, your secrets unfold,
A tale of science, forever untold.

    So let us marvel at your atomic dance,
And delve into realms of scientific chance.
Cesium, the element, forever we'll proclaim,
A symbol of progress, forever aflame.

SEVENTEEN

CESIUM'S BEAUTY

In the realm of elements, a force so bold,
Cesium, a power yet to be truly told.
With atomic weight, one hundred thirty one,
A metal so reactive, it shines like the sun.

In laboratories, it dances and gleams,
A catalyst for science, unlocking dreams.
Its reactivity, a marvel to behold,
Aiding in experiments, both new and old.

From flame tests to spectral lines,
Cesium unveils secrets, a tale that shines.
Its golden hue, a beacon of light,
Guiding scientists through the darkest night.

But beyond the lab, there lies another feat,
Cesium's timekeeping prowess, oh so sweet.

Atomic clocks, precise to the core,
Counting moments, forevermore.
    Tick by tick, it measures the hours,
A testament to Cesium's wondrous powers.
In perfect rhythm, it marks the days,
Guiding us through life's intricate maze.
    Cesium, a marvel of the periodic table,
A gem of the elements, both strong and stable.
From its atomic weight to its fiery glow,
Cesium's beauty, the world shall forever know.

EIGHTEEN

VIVID DREAM

In the realm of elements, a treasure lies,
A metal rare, with a gleam that defies,
Cesium, a name whispered on silent breeze,
Unveiling secrets that science seeks to seize.

From the depths of the Earth, it emerges bright,
A precious metal, casting a radiant light,
Its atomic dance, a symphony of grace,
Unraveling mysteries of time and space.

With a single electron, it rules supreme,
Unleashing power, like a waking dream,
Its fiery essence, a spark of pure might,
Igniting scientific flames, burning bright.

In atomic clocks, it finds its true home,
Ticking away, never ceasing to roam,

A guiding light, measuring moments in time,
Unveiling the universe's intricate rhyme.

    Cesium, a catalyst for progress and change,
A beacon of hope, in a world so strange,
Through its atomic structure, we aspire,
To unravel the secrets that fuel our desire.

    Oh, Cesium, element of wonder and worth,
A testament to the power of the Earth,
In laboratories, your beauty unfolds,
Unveiling the secrets that science beholds.

    So let us raise a toast to Cesium's might,
A symbol of knowledge, shining so bright,
In the realm of elements, you reign supreme,
Unlocking the mysteries, like a vivid dream.

NINETEEN

FOREVER SUBLIME

In the realm of scientific wonders, behold Cesium's grace,
A shimmering element with an ethereal embrace.
With atomic number fifty-five, it claims its noble throne,
Unveiling secrets of the universe, where knowledge is sown.

A symphony of electrons, spinning in delight,
Within its valence shell, a dance of cosmic light.
Its silvery luster, gleaming with celestial might,
In laboratories and stars, it ignites curiosity's flight.

Cesium, the timekeeper, with its atomic clock,
Unraveling the mysteries, like the ticking of a rock.
With precision and accuracy, it measures each second's toll,
Guiding explorers of time, towards horizons untold.

A catalyst for progress, it fuels innovation's fire,
Unlocking doors to new realms, where dreams aspire.
In spectral lines, it paints a canvas, vibrant and bold,
A testament to the beauty that science does behold.

Oh, Cesium, guardian of knowledge, oh so rare,
With your gentle touch, you move the world with care.
In laboratories and research halls, your presence does inspire,
To seek the truth, to push boundaries higher.

So let us celebrate this element, so grand,
As it dances with the cosmos, hand in hand.
For in its essence, beauty and wisdom entwine,
Cesium, the element, forever sublime.

TWENTY

INSPIRING

In the realm of science, a hidden gem does reside,
A metal so rare, with beauty deep inside.
Cesium, the element, with secrets to unfold,
A story untold, waiting to be told.

Its atomic number, fifty-five it claims,
A symbol of progress, igniting scientific flames.
Soft and silvery, it glimmers in the light,
A sight to behold, shining ever so bright.

Cesium, the catalyst of discovery's dance,
Unveiling mysteries with every single chance.
From its flame, a hue of blue so profound,
Unleashing knowledge, spreading it around.

Through spectrums and clocks, its magic does shine,
In timekeeping precision, a gift so divine.

Atomic clocks tick, with Cesium as their guide,
Guiding us forward, progress cannot hide.

With waves so precise, it measures our days,
Unraveling the universe in wondrous ways.
From Earth to the stars, it unlocks their secrets,
Cesium, the key to knowledge, it never retreats.

Symbol of innovation, a beacon of light,
Cesium inspires, guiding dreams to take flight.
In laboratories and classrooms, where knowledge is shared,
Cesium, the element, forever stands prepared.

So let us celebrate this element rare,
Cesium, the element, with wonders to share.
In scientific pursuit, it leads the way,
Forever inspiring, as we continue to sway.

TWENTY-ONE

KNOWLEDGE

In the heart of nature's realm, a wonder lies,
A gleaming element, beneath azure skies.
Cesium, the gift of scientific lore,
A tale of secrets, waiting to explore.

With atomic number fifty-five in its core,
Its electrons dance, a celestial score.
A golden hue, a metal bright,
Cesium's brilliance, a pure delight.

In laboratories, it takes its stand,
Unveiling mysteries, at science's command.
Its low melting point, a curious feat,
A liquid metal, so graceful and neat.

It whispers tales of spectral lines,
Guiding scientists, through cosmic signs.

Its flame, a vibrant shade of blue,
Revealing truths, both old and new.

    In atomic clocks, it finds its place,
A timekeeper, in the human race.
With utmost precision, it counts the ticks,
Guiding progress, with each moment it picks.

    Oh, Cesium, you measure time,
A constant companion, so sublime.
Through your essence, we seek to know,
The depths of knowledge, you help us sow.

    In laboratories and time's embrace,
Cesium, you leave us in awe and grace.
We celebrate your atomic might,
A beacon of knowledge, shining so bright.

TWENTY-TWO

SCIENTIFIC LOVE

In the realm of science, where secrets reside,
There's an element, rare and dignified.
Cesium, a beacon of light it shines,
Guiding progress with its atomic lines.

A metal so precious, its beauty profound,
A tale of discovery, so renowned.
In the lab of Bunsen, it first appeared,
Unveiling mysteries, so revered.

With its atomic clock, it measures time,
Unwavering pulses, so sublime.
A master of precision, it beats with grace,
Keeping the world in a rhythmic embrace.

Cesium, the element of innovation,
Igniting minds with its radiant creation.

A catalyst for change, it sparks the flame,
Unleashing ideas, no longer tame.

Through its atomic dance, it unveils the unknown,
Revealing the secrets that lie alone.
In labs and classrooms, its story unfolds,
Inspiring scientists, both young and old.

Oh, Cesium, in your atomic embrace,
We find solace, in your radiant grace.
A symbol of knowledge, you brightly shine,
Guiding us forward, through the realms of time.

So let us celebrate this remarkable element,
With wonder and awe, in every sentiment.
Cesium, a gift from the heavens above,
A testament to the power of scientific love.

TWENTY-THREE

FLAWLESS LAW

In the depths of the atom's core,
Where mysteries and wonders pour,
Lies a jewel, a precious gem,
Cesium, the timekeeper's emblem.

With steady rhythm, it does beat,
Guiding scientists in their feat,
A guardian of moments passed,
Cesium's touch, forever cast.

Its electrons dance and sway,
In perfect harmony they play,
A symphony of glowing light,
Illuminating the darkest night.

Cesium, a beacon in the dark,
A compass for the curious heart,
Unleashing secrets, unlocking doors,
Revealing what the universe stores.

Through cesium's lens, we dare to see,
The realms of possibility,
A world of knowledge, vast and wide,
Where innovation can reside.

Oh cesium, catalyst of change,
In every discovery, you arrange,
A symphony of progress and hope,
A testament to humanity's scope.

So let us honor this element true,
For all the wonders it helps us pursue,
Cesium, a source of endless awe,
A testament to nature's flawless law.

TWENTY-FOUR

EXPANDING HORIZONS

In the realm of atoms, a jewel does shine,
A metal that dances, a treasure divine.
Cesium, the catalyst of progress and change,
Unveiling the universe, a boundless range.

Within its core, a secret does lie,
A spark of brilliance, a celestial tie.
With atomic number fifty-five it appears,
A conductor of power, dissolving all fears.

A metal soft, like a whisper on breeze,
Caressing the senses, an elegant tease.
Its golden hue, a beacon of light,
Guiding the curious through the darkest night.

Cesium, the maestro of time's grand parade,
With oscillations precise, a rhythm cascade.

From atomic clocks, it measures the day,
A symphony of seconds, in perfect display.
    And in the lab, scientists find their delight,
Harnessing its power, unlocking insight.
From spectroscopy's dance, to lasers so bright,
Cesium unveils nature's secrets, day and night.
    Oh, wondrous Cesium, a muse to behold,
Inspiring the curious, the brave and the bold.
With every discovery, a new chapter begins,
Expanding horizons, where knowledge begins.

TWENTY-FIVE

GREATEST TREASURE

In the realm where time is measured,
There lies a secret yet untold.
A dancer in the cosmic ballet,
A treasure in the hands of old.

Cesium, the guardian of our clocks,
Unlocks the universe's mysteries.
With atomic grace, it marks the tocks,
Guiding us through eternal histories.

A single heartbeat, a fleeting second,
Cesium pulses, never to relent.
It counts the seconds, never beckoned,
In its presence, time finds its ascent.

Through the ages, it remains steadfast,
A constant, unwavering light.
A beacon in the darkness, unsurpassed,
Guiding scientists through the night.

Cesium, the catalyst of innovation,
Ignites the flames of discovery.
In laboratories, a revelation,
Its touch fuels scientific reverie.

From the depths of the unknown,
To the heights of celestial spheres.
Cesium's wisdom has shone,
Unveiling truths that calm our fears.

Oh, Cesium, a symbol of knowledge,
A testament to human endeavor.
In your presence, we seek homage,
For you are science's greatest treasure.

TWENTY-SIX

DAY BY DAY

In the realm where atoms dance and shimmer,
There lies an element, rare and glimmer.
Cesium, a metal, so serene and bright,
Unlocks the secrets of the cosmic night.

Within atomic clocks, its heartbeat thrives,
Ticking with precision, where time derives.
Its oscillations, a rhythm so pure,
Measure the seconds, forever endure.

Oh Cesium, conductor of time's grand show,
Guiding us forward, as moments bestow.
Through your atomic dance, we comprehend,
The vastness of space, where galaxies blend.

But Cesium, there's more to your story,
A flame, so vibrant, glowing in glory.
Burning with a hue, deep and intense,
A mesmerizing blue, captivating the sense.

You paint the night sky with ethereal light,
A beacon of wonder, through the darkest night.
And as your flames flicker, the world is aglow,
Revealing the mysteries we yearn to know.

Cesium, a catalyst for change and growth,
A symbol of knowledge, we all may boast.
In laboratories, where discoveries ignite,
You lead us towards progress, shining so bright.

So let us celebrate this element divine,
Cesium, the star in science's design.
From timekeeping to illuminating the way,
You unveil the universe, day by day.

TWENTY-SEVEN

CURIOSITY RESOUNDS

In the realm where time holds sway,
A silent guide, in hours it may play.
Cesium, the element of grace,
Measuring moments in a cosmic race.

With steady pulse, it counts the ticks,
A beacon of precision, it quickly clicks.
The cesium fountain, a marvel of might,
Unveiling the truth in spectral light.

From atoms danced, a symphony born,
A rhythm of waves, a knowledge adorned.
In laboratories, minds set ablaze,
With Cesium's secrets, they seek to raise.

A catalyst for progress, it does ignite,
The flames of innovation burning bright.

Unlocking doors to the unknown,
A journey of discovery, to call our own.
    Through time's embrace, we find our way,
Led by Cesium's unwavering sway.
A testament to human endeavor,
Revealing the mysteries we tirelessly gather.
    So, let us celebrate this element grand,
A symbol of knowledge, in our hands.
For Cesium's power, it knows no bounds,
In its presence, curiosity resounds.

TWENTY-EIGHT

MINDS HAVE GROWN

In the realm where secrets hide,
A shining beacon, Cesium's stride.
A catalyst of change profound,
Unveiling truths that were tightly wound.

    Radiant hues, a luminescent dance,
Illuminating the path of chance.
With atomic grace, it guides the way,
In the pursuit of knowledge, come what may.

    Unlocking doors to the unknown,
Unleashing minds that have grown.
From laboratories to distant skies,
Cesium's brilliance never denies.

    Tick-tock, the pendulum swings,
Measuring time with precise rings.

Its heartbeat echoes, steady and true,
A symphony of moments, old and new.
    Curiosity sparked, a flame ignited,
By Cesium's touch, our world enlightened.
A symbol of progress, of human endeavor,
It inspires, it guides, and it will forever.
    Oh, Cesium, element of grand design,
Your presence, a gift, so divine.
In the tapestry of science, you play your part,
Unveiling mysteries, igniting the spark.
    With each discovery, we stand in awe,
For Cesium, you are the eternal law.
A testament to our quest for the unknown,
In your radiant glow, our minds have grown.

TWENTY-NINE

UNRAVELING MYSTERIES

In the depths of time, a shimmering tale,
Resides a treasure, Cesium, so frail.
A lustrous metal, with secrets untold,
Unleashing wonders, as its story unfolds.

A beacon of science, a clock's steady hand,
Cesium, the master, of time's golden strand.
With atomic precision, it beats and it ticks,
Unlocking the mysteries, with its temporal tricks.

In laboratories, its dance begins,
Guiding researchers, through scientific sins.
A catalyst for progress, it sparkles and gleams,
Igniting innovation, in the realm of dreams.

Through spectrums it travels, a vibrant display,
Revealing the world, in a colorful array.

Its atomic glow, a mesmerizing sight,
Illuminating the unknown, in the darkest of night.

    Curiosity awakened, minds set ablaze,
Cesium, the muse, in captivating haze.
Its secrets unravel, like a symphony's score,
Inspiring wonder, forevermore.

    Oh, Cesium, element of grace,
A symbol of knowledge, in this cosmic space.
From timekeeping to discoveries profound,
You guide us forward, where knowledge is found.

    So let us celebrate, this element divine,
Cesium, the catalyst, in the march of time.
Forever we'll cherish, this gift so pure,
Unraveling mysteries, to forever endure.

THIRTY

FOREVER IN OUR HEARTS

In the realm of elements, a beacon of light,
Cesium, the wanderer, shining so bright.
Its atomic dance, a mesmerizing sight,
Guiding the curious through nature's vast might.

A single electron, so eager to roam,
Unleashing its energy, a powerful tome.
With each vibrant pulse, a secret it'll disclose,
Unlocking the mysteries that nobody knows.

From laboratory chambers to fields afar,
Cesium's presence, a constant shining star.
Fueling our quest for scientific discovery,
In its atomic embrace, we find the key.

Its flame burns fervently, in shades of blue,
Illuminating the night sky with a celestial hue.

With every burst of light, the universe unfolds,
Revealing its wonders, as stories are told.

    Cesium, the catalyst for progress and more,
A symbol of knowledge, forever to adore.
In this journey of discovery, it leads the way,
Guiding our footsteps, day after day.

    So let us embrace this element's grace,
As it paves the path for the human race.
Cesium, the spark of curiosity's flame,
Forever in our hearts, it shall remain.

THIRTY-ONE

INNOVATION'S PARTNER

In the realm of science, a marvel unfolds,
A catalyst for innovation, Cesium behold.
Symbol Cs, a knowledge's emblem, so true,
Let me paint a picture, a tale that's new.

Cesium, the spark that ignites the flame,
In laboratories, where breakthroughs aim.
Its electrons dance, in a valence so bright,
Unveiling mysteries, with scientific might.

A beacon of progress, it lights up the way,
Guiding us through the darkness, night and day.
In atomic clocks, it measures time's stride,
With precision and accuracy, it's our trusted guide.

Cesium, a messenger from the depths of space,
Revealing secrets of the universe's embrace.

Its spectral lines paint a celestial story,
Of stars and galaxies, in all their glory.
 Curiosity's companion, it fuels our quest,
Unraveling the cosmos, putting theories to the test.
From lasers to solar cells, its power we employ,
Harnessing its potential, with boundless joy.
 So let us celebrate, this element divine,
Cesium, the catalyst, that makes science shine.
A symbol of knowledge, a key to the unknown,
Innovation's partner, forever it is shown.

THIRTY-TWO

DARKEST NIGHT

In the depths of the unknown, it lies,
A gleaming element, in silence it sighs.
Cesium, the catalyst of progress and time,
Unveiling the secrets, so sublime.

Its atomic dance, a mesmerizing sight,
A beacon of discovery, shining so bright.
With each tick of the clock, time unfurls,
Cesium's rhythm, the heartbeat of the world.

From ancient sundials to atomic clocks,
Cesium's precision unlocks time's locks.
Its resonance resonates through the ages,
Guiding humanity, turning the pages.

Curiosity ignited, minds set ablaze,
Exploring the cosmos, in a celestial maze.

Cesium's light, a compass in the night,
Leading us forward, towards insight.

In laboratories, it dances with flame,
Revealing the mysteries, no longer tame.
Its spectral lines, a roadmap to explore,
The depths of knowledge, forevermore.

Oh, Cesium, element of grace,
In your presence, we find our place.
A symphony of atoms, a cosmic duet,
Forever grateful, we'll never forget.

For in your essence, we find our quest,
To unravel the universe, we are blessed.
Cesium, the harbinger of science's might,
Forever illuminating the darkest night.

THIRTY-THREE

CESIUM, THE ELEMENT

In the realm of elements, a star is born,
Cesium, the catalyst, the cosmos' adorn.
With gentle grace, it lights the way,
Unraveling mysteries, night and day.

Curiosity ignited, minds aflame,
Cesium whispers secrets, whispering its name.
From laboratory walls to distant skies,
It guides our quest, our yearning eyes.

A beacon of time, it measures the years,
Ticking away, erasing our fears.
In atomic clocks, its heartbeat true,
Cesium, the pulse that carries us through.

In science's embrace, it finds its solace,
Unveiling nature's laws, with every trace.

Through spectral lines, it paints the dance,
Of particles and waves, a cosmic romance.
  With spectral hands, it points the way,
To unseen realms, where dreams may sway.
Cesium, the key to knowledge untold,
Unlocking the universe, as it unfolds.
  Oh, Cesium, your presence profound,
In laboratories, you astound.
A symbol of progress, of innovation's might,
Forever guiding humanity's flight.
  In your atomic heart, we find our place,
A testament to curiosity's embrace.
Cesium, the element that sets us free,
To explore the depths of our destiny.

THIRTY-FOUR

SCIENTIFIC FLIGHT

Cesium, the catalyst of progress,
An element with power to impress.
In laboratories, its secrets revealed,
Unveiling wonders, yet to be unsealed.

A metal soft, a hue of gold,
Its atomic weight, a story untold.
With a single electron it donates,
A spark of promise, it perpetuates.

From clocks precision, it measures time,
Guiding us through life's intricate rhyme.
Its steady beat, a symphony of ticks,
A beacon of hours, it never tricks.

In the night sky, it casts its glow,
Painting stars with a vibrant show.
Flashing hues of blue and green,
A celestial dance, a mesmerizing scene.

In the cosmos vast, its presence known,
Fueling curiosity, it has grown.
From distant stars to celestial spheres,
Cesium whispers secrets in our ears.

A symbol of knowledge, a guiding light,
It fuels our quest, day and night.
Unraveling mysteries, unlocking doors,
Cesium, the element that forever soars.

With each discovery, a tale to tell,
Cesium's wonders, we cannot quell.
An element of wonder, a spark of delight,
Cesium, the catalyst of scientific flight.

THIRTY-FIVE

POETIC DELIGHT

In the realm of scientific wonder,
A shining star, a cosmic thunder.
Cesium, element of grace,
Ignites our minds, sets the pace.

A catalyst for progress, it does inspire,
Guiding us through knowledge's choir.
With spectral lines, secrets unfold,
Unraveling mysteries, stories untold.

In time's embrace, it finds its place,
Ticking with precision, never a trace.
A guardian of hours, minutes, and days,
Marking the path, in countless ways.

A beacon of light, it illuminates the night,
A constant companion, shining so bright.

Through the vast cosmos, it guides our sight,
Revealing the wonders, with its radiant might.
　In laboratories, its properties unfold,
Aiding discoveries, a story yet untold.
Symbol of innovation, a scientific creed,
Cesium, the key to progress we need.
　From Earth to the stars, it knows no bounds,
Unveiling the universe, with resounding sounds.
Oh, Cesium, element of awe and might,
Forever in our hearts, a poetic delight.

ABOUT THE AUTHOR

Walter the Educator is one of the pseudonyms for Walter Anderson. Formally educated in Chemistry, Business, and Education, he is an educator, an author, a diverse entrepreneur, and he is the son of a disabled war veteran. "Walter the Educator" shares his time between educating and creating. He holds interests and owns several creative projects that entertain, enlighten, enhance, and educate, hoping to inspire and motivate you.

Follow, find new works, and stay up to date with Walter the Educator™
at WaltertheEducator.com

www.ingramcontent.com/pod-product-compliance
Lightning Source LLC
LaVergne TN
LVHW051959060526
838201LV00059B/3729